To Shu + Sue

A SUITABLE STONE

with thanks

Steve

A SUITABLE STONE

*How geology has shaped the
British stones used for letter cutting
and fine carving*

·

STEVE GARRETT
AND
LIDA LOPES CARDOZO KINDERSLEY

CARDOZO KINDERSLEY · CAMBRIDGE 2024

Italic alphabet in green slate

First published by Cardozo Kindersley Editions 2024.

Photographs from the Cardozo Kindersley archive, Steve Garrett, Shutterstock library, Honister Slate Mine, Albion Stone, Adrian Nicholls, Stuart Vallis.

With special thanks to Fiona Boyd for beautiful illustrations that bring geology to life.

Typeset in Emilida, designed by Lida Lopes Cardozo Kindersley
and digitised by Eiichi Kono, Eben Sorkin & John Mawby Whistler.
Designed by Dale Tomlinson *(studio@daletomlinson.co.uk)*.
Printed in Germany by Ph. Reinheimer, Darmstadt.

ISBN 978-1-7395878-3-3

Copies of this book and other Cardozo Kindersley publications available from:
The Cardozo Kindersley Workshop, 152 Victoria Road, Cambridge CB4 3DZ
www.kindersleyworkshop.co.uk/shop

Front cover: testing a piece of rock in the quarry
Endpapers: Sedgwick Museum of Earth Sciences, University of Cambridge
(Reproduced with permission 2024)
Half-title page: a relief carving of a dove in Portland stone
Frontispiece: inscription carved in a local rock on the Isle of Bute

Contents

FOREWORD 6
Dr Nigel Woodcock

1 · INSPIRATION 9

2 · A MAP OF TIME 15

3 · MEET THE STONES 25

4 · DEEP WATERS 35
Welsh Slates

5 · VOLCANOES 49
Lake District Green Slates

6 · RIVERS AND LAKES 59
The Sandstones

7 · TROPICAL SHALLOW SEAS 69
The Limestones

8 · REFLECTION 79

THANKS 82
GEOLOGICAL GLOSSARY 83
BIBLIOGRAPHY 86

Foreword

It is a pleasure to introduce Steve and Lida's delightful book on the stones used for letter cutting and fine carving. This is engagingly written, beautifully illustrated and skilfully designed. The text and pictures combine to show the intimate, almost spiritual, relationship between the stone cutter and their stone, an association enriched by understanding the history of that stone through deep time.

Like Steve and Lida, I have my own close relationship with rocks and stone. I spent my career teaching and researching in the Earth Sciences Department of the University of Cambridge. I gained a good knowledge of British rocks and wrote textbooks and many articles describing and interpreting them. However, almost as a hobby, I began to research constructional stone in the many outstanding buildings in Cambridge.

What drew me to this esoteric subject? It was partly the challenge of identifying the stones with their poetic, mostly Anglo-Saxon names: Ancaster and Clipsham, Barnack and Ketton, King's Cliffe and Weldon; the main quarries in the Lincolnshire limestone belt that supplied most Cambridge stone. It was partly an interest in the routes that these distant stones took to Cambridge, by medieval inland waterways and inadequate roads in the Middle Ages, and by railways during Victorian times. But mainly it was the realisation that masons throughout the ages were seeking the most appropriate and most cost-effective stone for each project; they too were seeking A Suitable Stone.

However, the stone for construction had to meet less stringent standards than a stone for letter cutting. All the listed quarries supplied freestone, the mason's term for a rock that can be carved in any direction, at least for quoins and buttresses, plinths and string courses, door jambs and arches, and for window dressings: sills, jambs, mullions, tracery and head moulds. But only the finest Lincolnshire limestones meet the demands of the letter cutter.

Since publishing my Cambridge study, I have looked at the stone in over 120 south Cambridgeshire churches, but time constraints mean that I've had to ignore most monumental stone. So, apart from being charmed by medieval Purbeck marble, I've thought too little about the needs of the fine carver. It was therefore a delight to meet Lida and to reconnect with Steve. Together they have opened my eyes to their Suitable Stones. I hope that you, the reader, will get as much understanding and enjoyment from their book as I have.

Dr NIGEL WOODCOCK
Emeritus Reader, Earth Sciences, University of Cambridge

1 · INSPIRATION

We are told that we are in the midst of another age of extinction, which may prove to be humanity's last. What will we humans leave behind as evidence? We live in a digital age, with so much information available, but in a format that is transitory: we cannot be sure where it is stored, and there is no guarantee that future generations will have access to it.

Still, we want to leave proof that human beings can think, communicate with each other, be compassionate, and through this togetherness build a caring, better world. Cutting letters in stone is about recording and preserving ideas and memories of what is most important to us. Ever since the alphabet was invented, people have had the tools to inscribe what they had learned and what they felt, combining their creative talent and their innermost feelings with the very rock this planet is made of.

Stone not only comes from the Earth itself, it is also a material that will last into the future. For countless generations it has been used for inscriptions, and these will survive as a record for centuries to come.

Left: Cutting letters in situ in a York stone step

Many different natural and man-made surfaces can be cut into, but nothing compares with the quality and integrity of the stones of this Earth, which have proven their strength, their lasting qualities and their receptiveness to beautiful lettering. Some stones are better than others for letter cutting. A dense stone or slate will not let in water and will not weather dramatically; it holds the letter's edge and remains stable in colour. By contrast, a more

porous limestone will breathe in the damp air and allow lichen to grow on it; this stone may take the colour of its surroundings, and as it weathers the clean cuts will soften.

Universally, the traditional way of showing care for someone who has been lost is by memorialising their life with a marker, a gravestone. How can we letter cutters meet people's need to find a lasting focus for love? What can be said in a few precious words?

How will it look to the world? Which stone should we use? These questions take time and careful consideration, and are what brought Steve to the Cardozo Kindersley Workshop in Cambridge to commission gravestones on behalf of his family.

In the course of discussion, Steve the geologist and Lida the letter cutter recognised not only common professional interests, but also differences in approach as scientist and artist. A geologist's perspective is over millions of years, on a scale of continental land masses; whereas a letter cutter's knowledge is focused on a human timespan, dealing with the texture and qualities of each individual stone. Geologists use the word 'rock' to refer to the natural material in situ; whereas artists use the word 'stone' to mean a piece that has been worked by human hand. A neat but significant distinction. Their conversations lengthened.

Steve approaches what rock *is* from the analytical perspective of Earth Science: a knowledge of the formation, physical composition and strength of solid rock. Lida approaches what stone *does* from the associative perspective of a practical artist: a knowledge of how stone accommodates emotions and allows the right shape of words. We include first-hand descriptions of how each stone behaves under tools, written by the current members of the Kindersley Workshop.

On the one hand is the geologist's long view of the origin of earthly materials in 'deep time' spanning hundreds of millions of years, recording the impact of huge events and cycles in our Earth's history: deep oceans opening and closing, volcanic eruptions, pressures building as continents collide, mountains being eroded by rivers, sea levels rising as the land is flooded. On the other hand, the letter cutter's perspective is how we as temporary human residents choose to record our movements and memories, our transience on this Earth, by incising careful wording on selected small segments of a suitable stone: celebrating the miracle of its durability and opening another chapter in the story of the rocks we walk upon. A bond of shared knowledge grew between the authors, as recorded in this book.

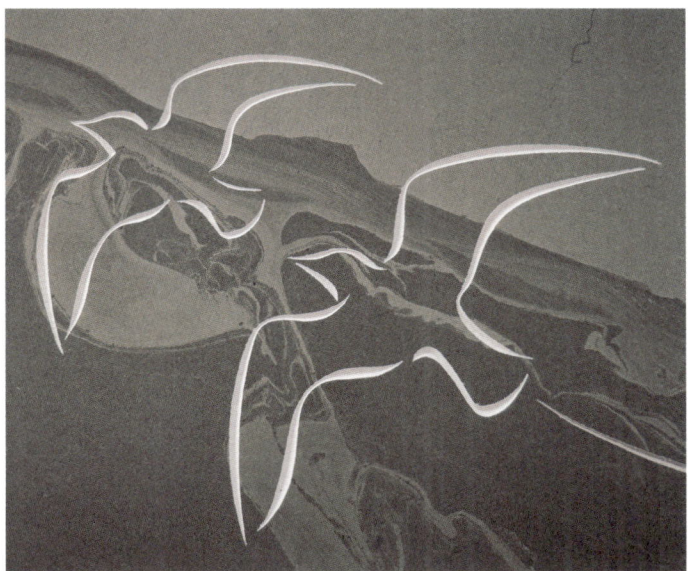

Seagulls soaring over characteristic green slate banding

2 · A MAP OF TIME

Steve: During a visit to the Workshop, I noticed a geological map of Britain on the wall. I asked a member of the Workshop what they knew about these strata, and whether they would be interested in learning more about the history of the stones that they carve. They answered: '*Yes please!*' Working in the present time with all kinds of stones, and knowing about their qualities under the chisel, the Workshop team was keen to learn something of the back story. So I gave a talk during my next visit, over an informal pizza lunch, to highlight the processes through geological time that lead to choices of suitable stones. We talked of the key skills that geologists learn through field observations and making maps by drawing.

To pay tribute to one of the seminal works of Earth Science, we later went to see an original version of the map published in 1815 by William Smith, the 'father of English geology', in the Sedgwick Museum in Cambridge. This iconic map of the southern part of Britain is now viewed as an object of veneration for geologists around the world. William Smith was a surveyor involved in excavating canals, which led him to observe the strata revealed, and in turn led to the map. As he cut through

William Smith's iconic 1815 geological map of part of Britain

layers of rock, he discovered the shape of the foundations of England. Smith's work in creating the map is beautifully described in Simon Winchester's *The Map that Changed the World*.

Lida: To me that map is a revelation in its portrayal of the lie of the land. It has a clarity comparable to the realisation of the structure of DNA. Beautifully drawn and coloured, the lettering too follows a long tradition of skilled printers and calligraphers, who always wrote the names of places so clearly.

As letter cutters, if possible we, like geologists, start our work by going to look at the site, to give us a feel of the place. We measure neighbouring features and see how the light falls. This gives us an idea of the best place to position a stone, and how to orientate it.

For gravestones it is not often possible to have free choice of positioning, as the grave is already there and usually looks east (except for that of the priest who has to guide the congregation and traditionally faces west).

The choice of material is much influenced by where the stone is to be placed. This does not necessarily mean that a gravestone in York for instance has to be made of York stone, but we do need to have that information at hand. That is why we have a geological map of Britain on our wall.

Steve: The geological maps of Britain we use today are remarkably similar to William Smith's map and are available to anyone online. So as we travel around

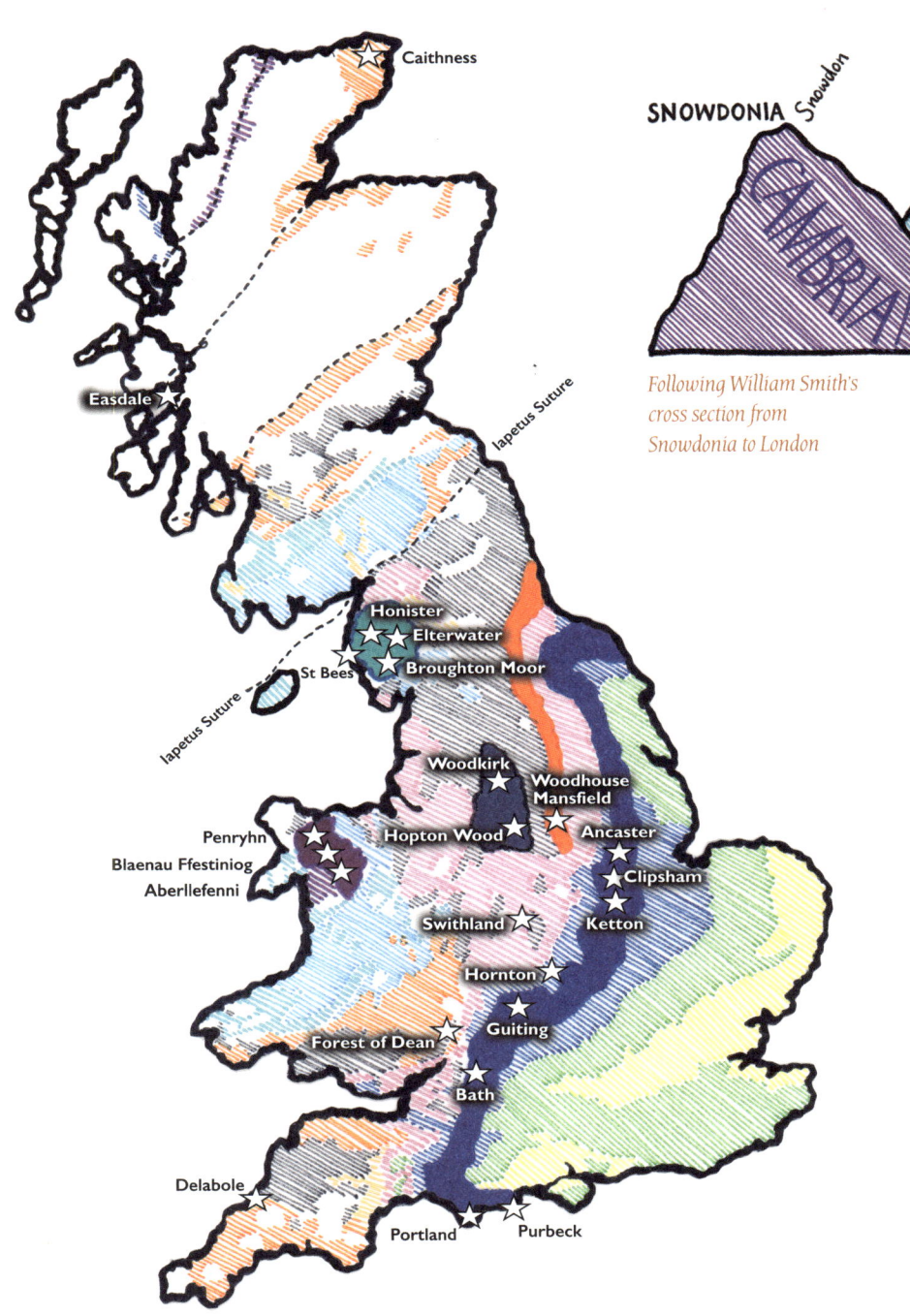

Following William Smith's cross section from Snowdonia to London

Today's geological map showing quarries of suitable stones

LONDON

Britain, we can find the age and composition of the rocks beneath our feet to help us understand the landscape around us. For example, when returning recently from Spain across the English Channel, the map explained the white chalk cliffs visible when approaching the coast – telling us we were nearly home.

One of the beautiful things about the geological map of Britain is that it shows how the ages of rock change across the land. In effect, it is a map of time. The youngest stones you use in the Workshop are of Jurassic age, which form a beautiful blue band on the map across England from Lincolnshire to Dorset. As we go towards the north-west, the stones you use get older: first, Carboniferous in central England; then Ordovician in the Lake District; and finally, Cambrian in Wales, the oldest stones commonly used in the Workshop. To the south-east are found younger Cretaceous chalk rocks which, although they form spectacular cliffs, are too soft and porous for your work. Younger Cenozoic sandstones and shales in the that region are even softer.

It is worth noting that Scotland is different geologically. Indeed, Scotland and England were on separate continents more than 450 million years ago. The ancient ocean between them closed over as the continents converged and formed mountains. The join or 'suture' runs from Lindisfarne to the Solway Firth, through the Isle of Man and then to Northern Ireland. The direction of the grain of rocks on the map changes on this boundary, from north/south in northern England to north-east/south-west in southern Scotland. The Southern Uplands, which represent the ancient Iapetus Ocean that separated the two land masses, stand up vertically, as they were squashed when the continents came together in what is called the Caledonian orogeny – an important event in the formation of the slate rocks we will discuss in later chapters. The join is called the Iapetus Suture and can be visited today on the shores of the Isle of Man.

Taken on a geological pilgrimage to the Iapetus Suture, Niarbyl Bay, Isle of Man

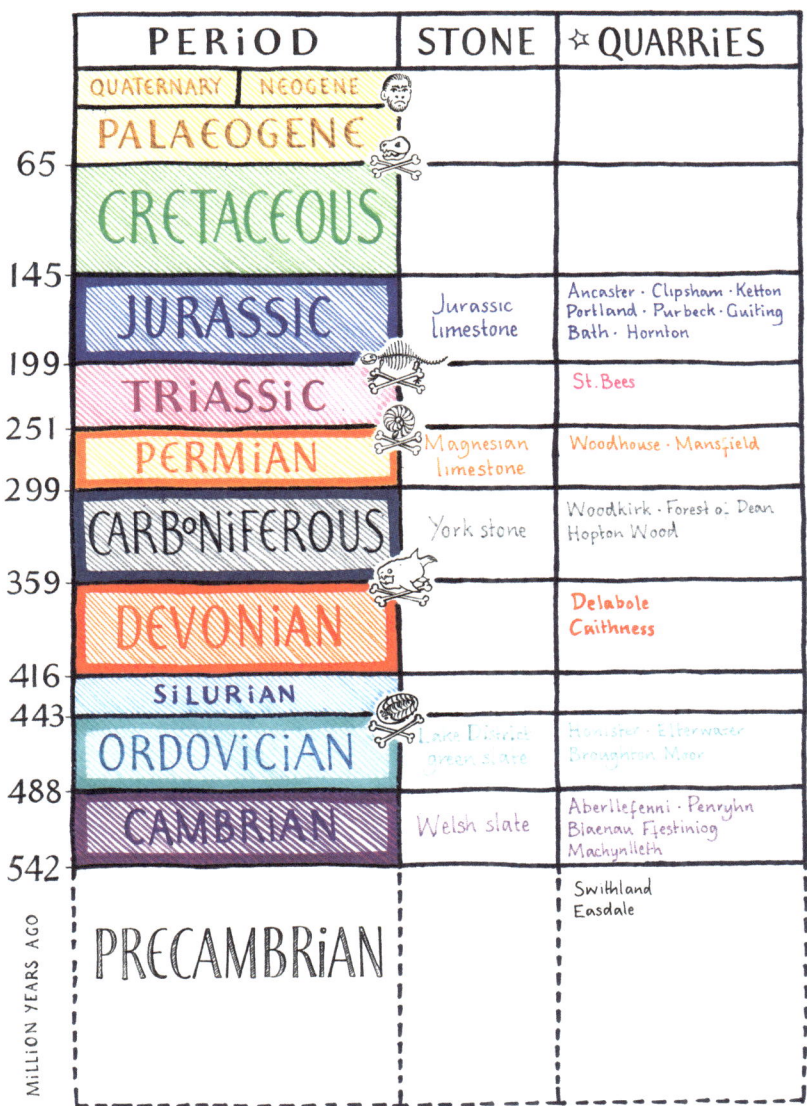

Geological time scale showing the greatest mass extinctions, suitable stones and quarries

Scotland and North America finally split apart when the North Atlantic opened around 60 million years ago – following the extinction of the dinosaurs. Lida, I heard you call this 'recently', so I am happy to see you're getting a perspective on deep time!

Lida: When you talked about our rocks being millions of years in the making, I felt rather humbled by explaining how long we take to cut our inscriptions. What is but a month or two in the magnitude of the creation of our basic material!

We cut stones to last for hundreds of years, maybe thousands. I am learning that they may even fossilise and our letters could still be around in a few million years, but the geological timescale makes our human measurement rather modest.

As letter cutters we pride ourselves on thinking long term, both backwards and forwards, but not quite as far back and ahead as you.

Steve: The Earth is 4.54 billion years old, as measured from the rate of radioactive decay of rocks. The materials which will last for future generations were forged over hundreds of millions of years by processes that have been active throughout Earth's history. Your suitable stones were formed between around 540 and 140 million years ago – and even that is a small window of time in the history of the Earth.

If the age of the Earth was twelve hours, creation of our suitable stones would take a little more than an hour (drawn on the Workshop clock face)

Lida: For most of us, the millions of years you mention are an unimaginable length of time. But even within this timescale, a lot of very good stones just don't behave in the right way for letter cutting: they can turn out to be hard as glass, or crumbly soft. So we have to narrow down our selection, considering the properties and qualities of different stones.

Smooth and dense.
Holds very fine detail,
for a long time.
Weathers slowly keeping
its dark colour.

CHAPTER 4
Welsh Slate

Hard and dense.
Holds detail in smooth or
riven surface.
Weathers slowly to
a softer colour.

CHAPTER 5
Lake District Green Slate

Gritty and hardwearing.
Good for paving,
masonry and big letters.
Weathers quickly to
a darker colour.

CHAPTER 6
Sandstones

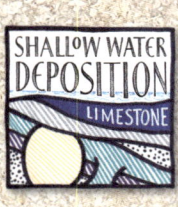

Soft and malleable.
Good for masonry, sculpture,
medium and big letters.
Weathers quickly into
its environment.

CHAPTER 7
Limestones

3 · MEET THE STONES

Lida: Our suitable stones for letter cutting are a subset of building stones produced by quarrying and mining. So the letter cutter is both benefitting from the Earth and actually exploiting it, tearing the rock from its bed and shaping it into the kind of stone we want it to be. In my 2023 book *Words Made Stone*, I compared the life of a stone, from bed to workshop and into the world, with a human life, from birth to fulfilment.

A quick calculation leads to an estimate that the amount of stone used over the life of our Workshop (almost 100 years so far) is only of the order of 350 tonnes – the weight of approximately ten lorry loads of road stone, which must be just a fraction of the daily traffic of this material on our roads today. I am proud that we have used such a small amount of stone with great sensitivity.

There are now fewer working quarries left in Britain that provide stone we can use. At the start of the 20th century there were around 3,000 different quarries operating; now there are only about 200. Of these, we use maybe twenty. Many of our stones are mainly used as

Left: Workshop samples of suitable stones

building materials but that world has changed dramatically. New materials have replaced real stone, and much is imported too. Big machinery has been introduced which is not practical or even possible to use in smaller quarries. Loss of expertise can also be a problem. However, there are times when small quarries re-open.

Hopton Wood stone, described by Ian Thomas in 2005 (see Bibliography), was much used in the first half of the 20th century by Eric Gill, Henry Moore and other leading sculptors and letter cutters. It works like marble, and can be polished to an attractive sheen. After a period of blasting, which made it useless to letter cutters, Unilever – the then owner – demanded in 1959 that hand quarrying be resumed for the renovation of their headquarters; this reversion in quarrying technique provided suitable material for letter cutting once again.

Apprentice piece in shelly Hopton Wood stone

Steve: Many of the working quarries today make their money from sources other than stone for building or letter cutting: supplying road stone, leisure activities involving wire ladders and rope bridges, extreme sports on tightropes or zip wires, or from the ultimate fate of being used for landfill. The Honister slate mine staff in the Lake District kindly hosted a visit in 2023 that made it clear that although quarrying continued, most of their income was from adventure tourism. At least these are ways of keeping the business viable and the quarries open, to provide more stones for letter cutting.

Modern uses for old quarries

Lida: In 1911, John Watson made a collection of all British building stones available in his time. This is now housed in the basement of the Sedgwick Museum of the University of Cambridge, where Robert Seidel of the museum staff showed us this comprehensive collection. Each one of the stones used by the Workshop is represented as a subset within the full spectrum of building stones. When I saw our suitable stones among these samples, there was something heartwarming about recognising them,

Suitable stones scattered across the John Watson collection

like meeting an old friend in a crowd. So our suitable stones are appropriate for building first, which makes them worth quarrying. We letter cutters are just a tiny appendix to the big world of building materials. However, we choose the best pieces for our work.

Steve: Nina Morgan wrote an introduction to gravestone geology in 2016 (see Bibliography) covering a wide range of materials. For this book, we will exclude from

discussion some of the stones that are not suitable for fine lettering, even though they may be widely used for other purposes. Softer rocks may be just too young: porous chalk is vulnerable outdoors to repeated onslaughts of wetting, drying and freezing. Igneous rocks such as granite contain large crystals which formed as molten magma and cooled slowly deep in the Earth; these can make wonderful building materials but may not be as good for letter cutting.

Lida: Granite is not our favourite material for cutting fine lettering, delicate flourishes or intricate carvings, but it is perfectly possible to chisel a good sharp edge. Most British granites have dominant differing colours which do not help our letters stand out – so I often refer to granite as 'glass salami'.

Carving granite needs muscle and a heavy hammer

Steve: This is because of the size of the crystals, reflecting the slow cooling of the magma, the variability in colour and hardness reflecting the different crystals (mainly quartz, feldspar and mica), and the toughness of the stone itself. I saw a monument in Cumbria with letters carved in a coarse-grained igneous stone where the visual texture was so dominant that the words were illegible – a great example of what not to do!

So our suitable stones are a 'Goldilocks problem': they must be not too hard, not too soft, not too porous, not too granular, not too complex…

An unsuitable stone. The letters in granite, at Alston, Cumbria, are virtually illegible

Lida: Our database of a century of stone used in the Workshop suggests that the most common British stones we have chosen for our work are Welsh slate, then limestone, then Lake District green slate, then sandstone.

Steve: The geological convention in storytelling – for geologists spin tales lasting millions of years – is that we start with the oldest rocks first and work our way forward in geological time. So let's start with the Welsh slates, which were originally deposited in deep seas. Then we will turn to the Lake District slates, which are the result of volcanic eruptions. Rivers and lakes gave us sandstones. Finally, tropical shallow seas gave rise to limestones. These can be viewed as representing parts of the 'mega' cycles of ocean formation and mountain building which are a key part of the life of the Earth.

4 · DEEP WATERS
Welsh Slates

Lida: If I had divine power, I would have laid down more slate for letter cutting, as it is perfect for this purpose. It shows contrast. It is dense. It keeps its edge. It is even in colour. A letter cutter looks for stones that invite the chisel into the stone – welcoming us in. Indeed, the first book David Kindersley and I wrote in 1980 was titled *Letters Slate Cut*, extolling these qualities and showing examples.

Welsh slate Roman capital letters with original depositional banding (bottom left to top right)

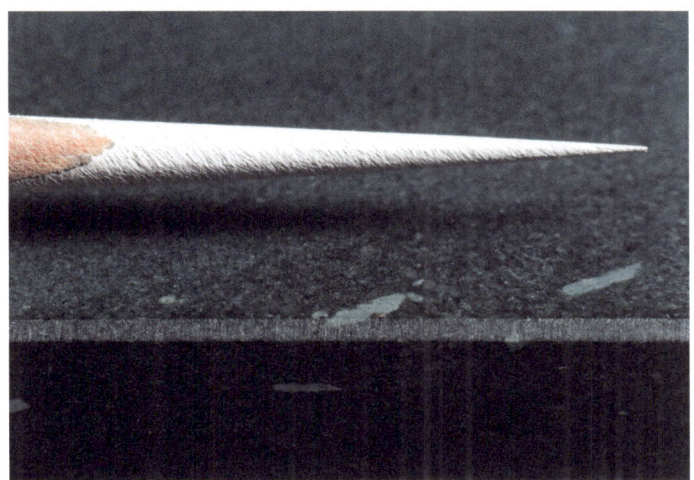

A sharp pencil gives a precise line

So slate is the most wonderful material for cutting letters. It is also the best surface when we first draw the design with a white pencil – and for rubbing out with a damp cloth. It gives a clean line during cutting, with each letter clearly visible as the light dust provides a contrast with the dark slate. It is tight-grained, so you can cut even very small letters; each tiny serif shows up precisely, and of course every little wobble will show up too, meaning that the skill of the letter cutter is on full display.

Some of the first questions when choosing a stone are: Do you want a dark or light stone? Do you want a stone that weathers? Welsh slate's blue-grey colour goes darker when wet. It is impervious so it is slow to weather and resilient to lichen.

Formation of slaty cleavage which cuts across the depositional layers

Steve: The special character of Welsh slate was documented by Terry Hughes and others in 2016 (see Bibliography), leading to its recognition as a Global Heritage Stone Resource by the International Union of Geological Sciences (IUGS).

The fine grain of the rock reflects its slow deposition as mud in deep water in ancient oceans. The dark shade reflects an environment lacking in oxygen. The Welsh slates are from the Cambrian period, which was characterised by the increasing appearance of fossils as life exploded. The original bedding can sometimes be seen as very subtle banding across the stone, or in the traces of beasties that burrowed into the mud in search of food.

These old rocks have enjoyed significant deformation and compression by late tectonic movements and chemical alteration. These changes happened during the formation of the Caledonian mountain belt in Early Devonian time (about 400 million years ago), when the Iapetus Ocean between Scotland and the rest of Britain closed. Scottish slate quarries (such as Easdale) concentrate around the major faults which originated at this time. Similarly, further south in Cornwall, the major Delabole slate quarry lies very close to a separate boundary between colliding landmasses identified in 2018 by Arjan Dijkstra and Callum Hatch (see Bibliography).

Cutting in the upright position – looks after our backs and lets the dust fall down

As well as geological age, the original chemical composition and subsequent changes in constituent minerals also have a big effect on the hardness of rock. So these slates are no longer soft mud or fissile shale. The effect was the reforming and realignment of the mud and clay particles with different, stronger compositions, so that the fissures are in a different direction to the bedding; we call this 'slaty cleavage'.

Lida: What interesting terms you geologists use! We have our own terms and would call the banding in our stone 'striations'. Letter cutters are also reforming the slate in our own heat of the moment with the force of the hammer, guiding the chisel into the stone while not wanting to barge in and break the stone aggressively. Welsh slates are very kind to the chisel, so we don't have to sharpen our tools very often. In the past, chisels had to be toughened, but now we can use tungsten carbide tools that allow us to work more efficiently.

As slate is soft, it makes more of a 'thud' than a 'ping' when cut. It is tempting to hit it hard and take too much out in one go, but if you do, it chips easily. We need to go gently, or the material shows marks and chips. We must be kind to it and treat it with respect.

Steve: This reflects the fine grain and consistent composition of the slate minerals. There is little or no hard quartz in these deep water deposits, as there is in granite or sandstones. The deep water of deposition was a long way from the land where quartz sand grains originate. So this makes the stone easier to work with. Later chemical processes can also lead to the growth of cubic iron pyrites ('fool's gold') crystals, which can be cut through, although with a feeling different from the rock matrix within which they sit.

The hardness of minerals and rocks is measured by Moh's hardness scale, 1–10. Quartz has a value of 7, contributing to the hardness of granite and sandstone; slate has a value of 5.5; iron pyrites 6, meaning resilient yet workable.

Lida: There are many variations of the blue-grey Welsh slates: tough roofing material from Penrhyn, or the softer slates we prefer using from near the town of Blaenau Ffestiniog, where a few quarries are still active. In the past we used a lot of slate from Machynlleth, but since the quarry closed, we can only get some small pieces.

In 1980 we used a Swithland slate from Leicestershire for the original grave marker, or ledger stone, of Richard III. The magnificent slab had been in the centre aisle of the cathedral for centuries, and at last was given a worthy use. It is now displayed in the King Richard III Visitor Centre, as a new marble memorial in marble was made for the re-internment of the king's bones in 2015.

Welsh slate is my favourite stone for cutting letters, as you can achieve such a crisp edge and it's a very reliable and predictable material. It is also good for carving, as you can execute a finish with lots of fine detail, although it takes more time to cut than a softer stone like Portland. I think it would be fair to say that Welsh slate is the favourite for most of us in the Workshop.

Recently I have been working on a Welsh slate sundial for Emmanuel College, Cambridge. The piece has some subtle banding which adds life to the stone. I prefer this to a completely blank canvas. When cutting, there is a high contrast between the darkness of the stone and the lighter dust left behind from the stone taken away. I find it doesn't make my chisel go blunt too quickly, so I have time to cut a fair amount uninterrupted. Because the stone isn't porous, I can 'flood paint' the surface and then rub the excess paint away with wet and dry paper, resulting in a very neat finish. It is also slow to weather. The marks I have made will last for many, many years.

EMILY BUNTON

We had to sit on the floor for three weeks to carve this stone

Steve: Swithland slates are Charnwood rocks – the oldest rocks in England and Wales from the Precambrian era, more than 700 million years ago. So, perhaps surprisingly, the oldest rocks are not the hardest, even within the same broad rock type. Indeed, the Lake District green slates are harder, reflecting their fiery volcanic origin.

Lida: There are blue-grey slates in Cumbria and Scotland too. Although they tend to be harder and more brittle, more chippy, this does not stop us from using them.

The other, ultimate reason for using slate is its weathering properties. It is used traditionally as waterproofing, as it is impervious; so it holds its shape and keeps out the water. For masonry and huge letters, my first choice would not be slate: the cleavage prevents it from breaking away reliably, and if we were to hit it too hard it risks shattering. We often ask the quarry to

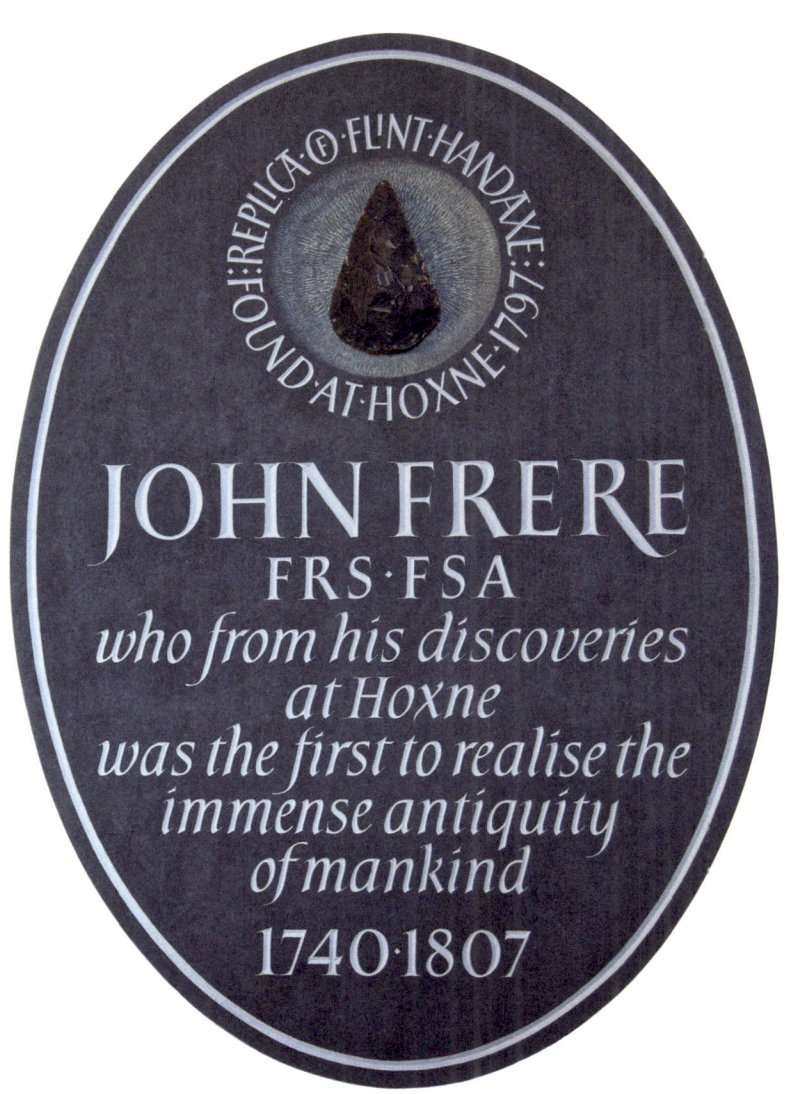

Welsh slate memorial to John Frere in Finningham church, Suffolk

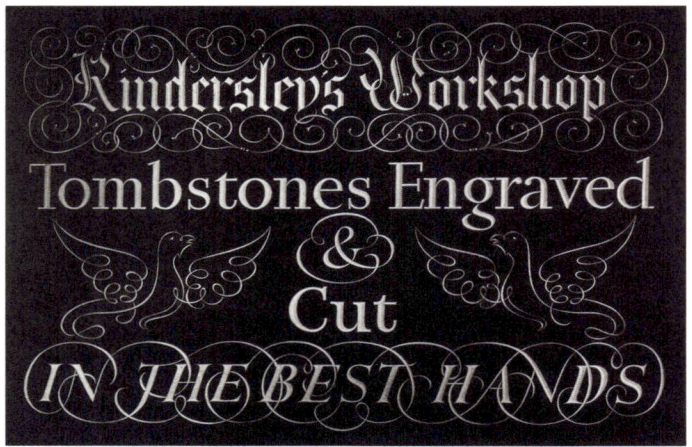

A pun on the 'Baskerville slate' in Birmingham Library – fine flourishes engraved with a burin (a handheld steel tool used for engraving in metal or wood)

shape the top of a headstone for us, as they have the right equipment to do it more efficiently than by hand. When we undertake masonry in slate by hand, we must work slowly and carefully, leading to higher cost.

Steve: Going back to our geological mega cycles, the Welsh slates represent deep oceans which then closed over as continents converged. The resulting volcanic eruptions formed our next suitable stones.

5 · VOLCANOES
Lake District Green Slates

Steve: Lake District green slates were originally deposited from clouds of hot ash with small rock particles ejected from volcanoes during the Ordovician period (about 455 million years ago). Geologists call these welded tuffs, or ignimbrites. Imagine the tumult as these materials were erupted with force from within the Earth, and then the incredible forces as they were compressed during the Caledonian orogeny to develop slaty cleavage. The green hue comes from minerals resulting from chemical alteration, such as chlorite.

Lida: The attraction of green slate is that it has both colour and complexity and looks more natural in the landscape. Within the Workshop collection of samples, we get excited about the variety of textures, densities

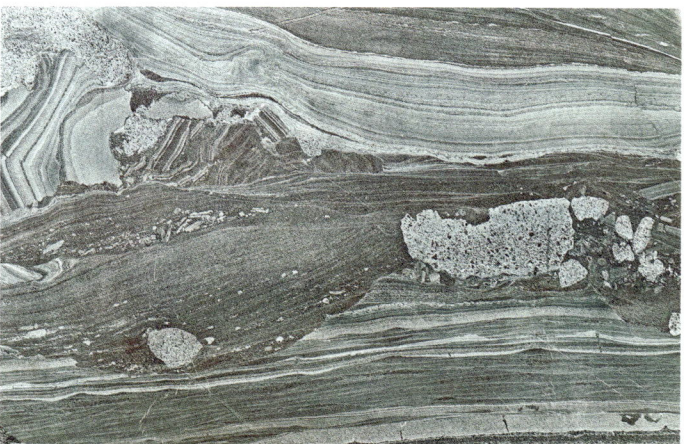

Layers from successive eruptions – later disrupted by major events

and colour changes in pieces that come out of the Lake District quarries. They have beautiful banding where the volcanic ash has fallen to form coarse layers at the base of the deposit graduating into finer material at the top, from light sea green to a dark and moody shade forming ripples like ocean waves.

Steve: The green slates from the Lake District are special to me as they are used for my family's headstones. Visiting the Honister and Elterwater quarries enabled me to get a sense of where they originated.

Honister Slate Mine, Lake District

Elterwater Quarry, Lake District

Left: Honister Slate Mine – the ceiling is a fault plane forming an upper limit to the slate

Lida: Elterwater slate is hard to find, as the quarry has closed. We now use Broughton Moor instead, which comes from the other side of the Old Man of Coniston and is more grainy. The other Cumbrian slate we use a lot is Honister, which is very clean, a little more brittle and a very dark green with characteristic volcanic textures. This slate is as dense as the Welsh slates, but tougher to cut. You really have to hit it to make a mark, and it gives a wonderful ring as you work it, as in my son Vincent's haiku:

> You can hit it hard
> The smooth surface of green slate
> New marks in ancient
>
> VINCENT KINDERSLEY

Vincent Kindersley cutting an inscription in green slate for Kettle's Yard, Cambridge

'I find it tricky to begin cutting green slate, as the chisel has to be very sharp and it feels as if it is going to slip. When starting with a serif, we need to tread carefully. It feels gritty and it is quite a tough stone so the chisel needs sharpening a lot. I do like it, but I sometimes find it frustrating as you have to work harder to get a clean edge than with say Welsh slate. It is a darkish colour and the cut turns into lighter shades. There is something light about it in a churchyard. I like the layers and other bits in it. It is rewarding, though hard work.'

STUART VALLIS

The ringing sound of Lake District slate has been used to make music: the lithophone which is on display at Brantwood, John Ruskin's house near Coniston, was made as part of a project in which stone instruments were played by Dame Evelyn Glennie in 2010.

Green slate 'lithophone' at John Ruskin's home, Brantwood, Lake District

Steve: In some of the rocks there are small fragments from previous eruptions that have been ripped up and carried along in the ash cloud. There are also subsequent faults which cut across and deform those layers, perhaps as the deposits slumped a little under gravity early in their life. These textures give us a lot to look at and tell the story of a violent creation that we would not want to have been anywhere near at the time.

Lida: We stone carvers are also using force, but on a very modest scale. We have two distinct ways of dislodging a piece of material. The more effective way is to stab into the surface, which controls most of the chipping. Whereas with softer slates we can chase along the length of the cut, travelling by chisel from the bottom to the top of the letter.

Stabbing straight into the layered slate

Steve: Welsh slate and Lake District green slate were compared by students of geology at the University of Manchester in 2017, using a variety of laboratory measurements. Roland Thompson at Honister quarry introduced us to some of their Masters' dissertations, which established that the physical and chemical properties of Honister slate makes it measurably more durable.

Lida: Green slate does change with weathering, as its minerals alter. Its colour can soften with time, becoming almost orangey. Lichen can grow on it, but not in it. Sometimes people feel the need to clean it: however, they should only use water and a washing up brush.

Apart from choosing it for its beauty and lasting quality, people also choose it for their love of the Lake District.

Steve: For example, Hawkshead churchyard in the Lake District is full of this stone, and it does look beautiful en masse.

Beautiful light on green slate headstones. Hawkshead, Lake District

After the volcanic eruptions in the Lake District, the next phase of our geological mega cycle, the oceans closed, continents collided, granitic mountains formed and were then eroded to provide the material for sandstones: our next chapter.

6 · RIVERS AND LAKES
The Sandstones

Steve: When Scotland and England collided during the Caledonian orogeny, mountains and highlands were forced upward. These were then eroded, with mineral grains swept away by fast-flowing rivers and deposited downstream in river beds, lakes and deltas going out into the sea. These processes reshaped the raw material of highland granite, providing finer-grained material which was deposited downstream. The resulting sandstones are largely composed of quartz grains and flat mineral grains such as mica. When the sandstones are fine-grained and consistent in texture, with the mica grains aligned along bedding planes, the rock parts very nicely along its bedding and might be suitable for letter cutting.

Lida: Sandstone is great for masonry, paving and cutting large letters. Because of its grainy and rather loose texture, it breaks away reliably; but it isn't really suitable for cutting small letters. Once installed, the front layers can flake off as the water seeps in along the bedding layers.

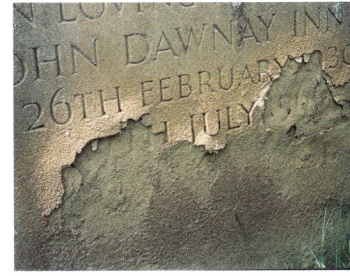

If water penetrates sandstone layers, surfaces peel off over time

The general feeling in the Workshop is that it eats chisels and weathers quickly, softening the definition in the letters.

Some sandstones gain a beautiful patina with time, and some become 'dirty' looking, although it is difficult to be general about this stone as it comes in such a wide variety of colours and textures.

Steve: Carboniferous 'York stone' is a generic term for fine-grained sandstones from various sources, such as river deposits, within the Coal Measures and Millstone Grit of northern England. There are layers of sandstone,

York stone showing black flecks (the letters are painted)

The freshly cut stone… …when weathered

shale and coal, as river systems travelled across the landscape and swampy forests grew which subsequently turned into coal. It is striking how pure the sandstones are: they don't seem to include much coal material, rather cutting across the coal seams, which made tracing those seams in mines complicated. These patterns repeat in layers which reflect the migration of river systems across extensive flood plains.

Lida: The broad term 'York stone' confusingly now includes stone imported from Asia, which can be different in its cutting quality. Recently one piece of York stone paving took three times longer to cut than expected. It turned out that this stone (called 'Haworth Heath') was imported. Misleading and very annoying. This Indian sandstone was glass hard and we broke a chisel with our first blow and discovered what we were up against. Our business involves making dust, and we are aware of the risks involved; this doesn't prejudice the choice of stone – but it does affect our approach. Silica content in the stone is something that carvers are now much more aware of.

One example of using real York stone is a memorial paving stone for Stephen Hawking at Gonville and Caius College, Cambridge, installed in 2018. This celebrates an equation describing the entropy of black holes. A flagstone was chosen to fit in with the existing paving, with the encouragement to *'Remember to look up at the stars and not down at your feet'.* Another example was a large plinth for a statue of Charles Darwin at Shrewsbury School with bold, clear letters – such a joy to cut!

Steve: The challenging attributes you refer to reflect the mineral composition. These quartz grains when cut create a safety hazard from silica dust. The texture of a flagstone can also reflect the layering of mica particles. The resistance to chisels is because these hard sand grains are composed of quartz eroded from granite mountains.

Caithness stone from Scotland is a much finer sand than York stone, as it was deposited in lower energy environments, in lakes during the Devonian period. This was an interesting time in geological history, as the first land plants emerged and were preserved in minerals

Cutting is solitary work…
…but everybody can join in with painting

Fine detail can be achieved in York stone, but it is perfect for big bold letters

from hot springs in the Rhynie Chert, north of my home in Aberdeenshire. I was pleased to come across a couple of letter cutters in St Andrews inscribing names into Caithness stone pavement slabs outside a university building.

The York stone I've just finished cutting is even in texture with a warm colour but it is an abrasive material to work with. After drawing a straight line onto the stone, the fine point pencil immediately required re-sharpening. The same applied to the chisel, which meant that frequent trips to the sharpening blocks were needed. There is a short period in which the chisel cuts with intent but it soon loses its edge and feels as if it is slipping over the surface rather than cutting into it.

The letter height for this headstone varied from 22mm italics to 50mm upright capitals. Even though this type of material demands a deeper cut letter, in flat lighting conditions the lettering can look slightly lost. When the sun shines, however, and shadows are created the letters have a definite but subtle presence.

STUART VALLIS

The first 'D' is unpainted, and the second one is painted a darker colour to make it show up

Sometimes we put a bit more varnish to make our oil-based paints more lasting

Lida: There are good reasons for selecting Caithness stone. It is fine grained and cuts as sharply as slate. It also looks quite dark when freshly quarried and when finely rubbed it can go even darker; then when it weathers, it goes a light sandy colour and the inside of the lettering can even go a little blue.

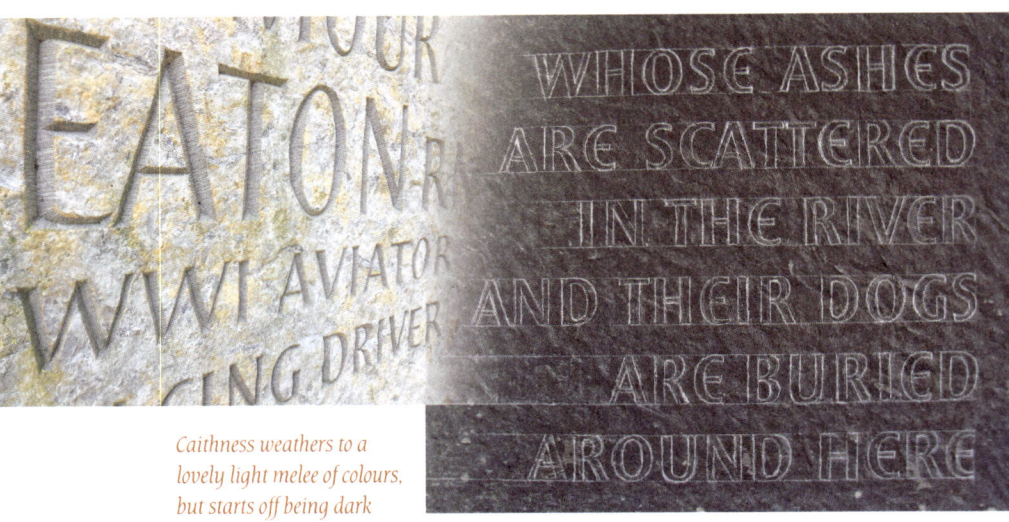

Caithness weathers to a lovely light melee of colours, but starts off being dark

Steve: Returning to our geological mega cycle, following erosion of the mountains, tens of millions of years later the seas encroached upon the land and limestones were deposited – our final group of suitable stones.

7 · TROPICAL SHALLOW SEAS
The Limestones

Steve: After the immense geological upheavals of throwing up mountains and massive erosion by rivers, warm tropical seas flooded the land and sea levels rose in Jurassic times. Because of continental drift, Britain was then positioned further south than today – at the latitude of what is now the Caribbean Sea or the Indian Ocean. These warm, tranquil seas were a fertile place for life, such as shelled organisms, and for the chemical deposition of calcium carbonate from living things. The resulting limestone rocks contained a mixture of carbonate mud, sand-sized carbonate spheres called ooids, and broken-up pieces of marine shells, all cemented together by crystalline calcite.

Lida: We use a lot of different limestones with a great variety of 'shelliness', hardness, 'crumbliness' and texture. They give a very sharp cut and they weather well as they exude their own protective calcium carbonate shell. Over time this takes on the colour of the environment: blue by the sea, green under a tree, dirty grey in a city, or black in a metropolis – though less so now that we are more aware of pollution and are doing something about it.

I was recently working on a memorial in Portland stone. This particular piece was to be laid flat outside and unpainted. When working in this stone it becomes second nature as a letter cutter to make sure the letters are deep and strong, even more so in an outdoor context with the limestone battling the elements. My aim was to push the lettering towards a bolder letterform, with a lower contrast between the thick and thin strokes, as well as making sure they were as deep as I dared.

Ordinarily we refine the final letterform using a 'chasing' technique. Here your chisel is typically pointing in the direction of travel, leaving perpendicular tool marks, a position which consequently limits your depth. However, another technique called 'chopping', can be used to remove large amounts of the initial material, requiring the chisel to be held at a much steeper angle; by adopting this method and becoming increasingly more considered as you near the letters' final shape you can achieve a much greater depth than otherwise. It is this relationship between the type of stone, the letterform and the way we choose to cut it that fascinates me. I really enjoy approaching a new stone, sitting down to start a new inscription and thinking 'now, what does this stone require from me?'

JOHN MAWBY WHISTLER

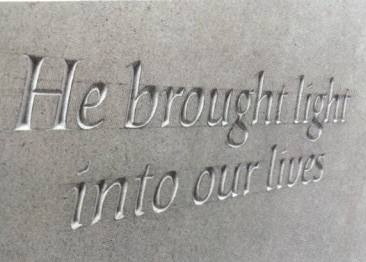

Left: beautiful depositional layers (Portland Quarry, Dorset)

I'm currently cutting one of four house numbers in Ancaster stone adorned with lettering and fleur-de-lis relief carvings. It is not a stone that we work with very frequently as it's not strong enough in its structure to make a lasting headstone. It's favoured as a building stone and once set into a wall it is well protected from weathering and other hazards. As a limestone Ancaster is relatively soft to cut, with a warm, slightly sandy colour and quite a lively texture. It can be a little crumbly, and achieving a sharp edge in it can be a little more difficult than in denser limestones such as Portland. There is a softness which makes it easier to work and faster to cut, especially on a large scale like that of the recently built wall at Emmanuel College, Cambridge. You can achieve results fairly quickly while cutting and the chisel isn't impacted too greatly, so it doesn't need much sharpening. Relief carving in this stone is possible at a good size, but fine detail can be a challenge. The cut surface is of very low contrast in appearance due to the pale dust against a light-coloured surface. So good lighting is necessary to see clearly while cutting and a deep cut is essential for lasting visibility. This is the reason why we often paint lettering and carvings in Ancaster stone. I like the colouring very much and I think this stone makes a lovely and unintrusive canvas for painted letters.

FIONA BOYD

Cutting the inscription at Emmanuel College, Cambridge, is a Workshop effort

As limestones are more open than the slates, they grow lichen that can be colourful and enhance their weathering, although some lichen can be very invasive and destructive. Limestone is often chosen for gravestones as it is light in colour and it weathers gently in its setting.

Steve: There are layers of coarse- and fine-grained material in limestones. Unlike the layers of the sandstones, which are driven by sideways migration of river systems, those in limestone are caused by vertical sea level changes. When the sea level is high and land is flooded with calm water, fine-grained material is deposited which will likely make a good stone for letter cutting. When the sea level drops, rough waves cut into shelly reefs and deposit large shell debris downslope. The resulting stone can look spectacular, but is likely to be more challenging for letter cutters.

Lida: Portland roach stone is one of the 'rough wave' materials you mention, and often contains the inside of shells, open and empty, which makes it tricky to cut in fine detail. Portland stone is very fine and one of the limestones we use most.

Steve: Portland Stone was recognised as another IUGS Global Heritage Stone Resource following a paper of 2013 by Terry Hughes and others (see Bibliography); and Gill Hackman's book of 2014 documents Portland stone's critical role in the buildings of London. However other oolitic limestones like Purbeck 'grub' and Hopton Wood stone have natural cement inside the shells, so letters can be cut more consistently. Then there are stones from Lincolnshire, such as Ancaster and Clipsham which can have a blue and fawn colour. The colour changes are due to changes in cement during and after the burial of the rock – complex chemical processes called 'diagenesis'. In a way this could be thought of as 'cement mixing' of different ingredients on a grand scale over geological time.

The material can change completely. This starts with alteration of the living shells which were created in aragonite; with burial they become fossilised in a more permanent mineral calcite. This can happen fairly early in the history of the rock, before it is buried and pressurised, as seen in the undistorted

Diagenesis changing the colour of Clipsham stone

shape of the fossils in Purbeck 'marble'. Diagenesis can alter the composition of the rock dramatically. Magnesian limestone (Permian) has had the calcium carbonate replaced by magnesium carbonate.

Lida: One of the very first things I carved in the Workshop in 1976 was a quote from Keats that David Kindersley had drawn onto paper. I translated this onto Woodhouse yellow Magnesian limestone which is lovely and soft to cut.

'Lamia' by Keats, drawn by David Kindersley, permanently hanging on the Workshop wall

When we went back to the quarry for more pieces, we sadly found that they had closed it and redeveloped the site. Maybe in the distant future it will be rediscovered and reopened.

An alternative stone with a similar yellow colour is Ketton. It is even more open-textured, with small oolite balls and a crumbly texture which is satisfactory for building and big lettering, but makes smaller detail more challenging to cut.

Steve: The oolite balls or spherules are grains of calcium carbonate that grew chemically in the sea as waves and tides washed them back and forth on the floor of shallow warm seas. The softer nature of the Ketton stone reflects less chemical action, so it doesn't have a strong cement matrix holding it all together. These oolitic stones are very commonly used as building stones across England. The oolite can look porous and spongy, or like bone.

You told me that you had two Ancaster gravestones break at ground level as the stone let in the damp that then froze. So limestones can be 'too porous' – the Goldilocks problem again. An open, porous matrix will allow water in; when it expands in summer and freezes in winter, this can crack the stone. Diagenesis and chemical cementation during burial in the Earth is the key to filling the open pore spaces with a solid matrix, providing a consistent stone for letter cutting, as seen in Hopton Wood or Portland stone.

Lida: In the past, when villages had their own mason and small quarry, only local stone was used. But there is no stone in Cambridgeshire so it has always been brought in from all around.

Steve: Cambridgeshire only has clunch, a form of soft chalk. Nigel Woodcock and Euan Furness (see Bibliography) estimated in 2021 that over two-thirds of stone used in buildings in Cambridgeshire is limestone from Lincolnshire. The variety of stone used in Cambridge was well documented in Donovan Purcell's book *Cambridge Stone* (1967).

Lida: There are many wonderful Irish stones too, like Kilkenny 'marble' which is a beautiful material, full of character and very demanding.

Steve: This is a fine-grained dark limestone deposited in deeper water, mention of which completes our journey through time and our geological mega cycle. We have travelled full circle back to deep water, where we began.

Soft enough to shape, but fine and dense enough to hold minute detail; when a piece of work asks for both fine carving and lettering, Hopton Wood must step up and take a bow. I was incredibly glad for it to be the stone of choice for the memorial and portrait of the poet Geoffrey Hill in 2021. This plaque's design asked for everything – fine, flourishing italic lettering, strong, classic capital lettering for naming, a moulding to carve for the edge and, most challenging of all, a profile portrait of the poet in bas-relief.

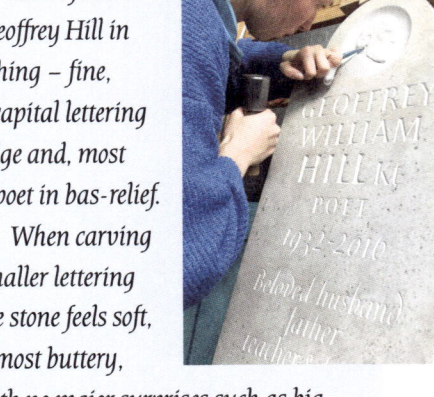

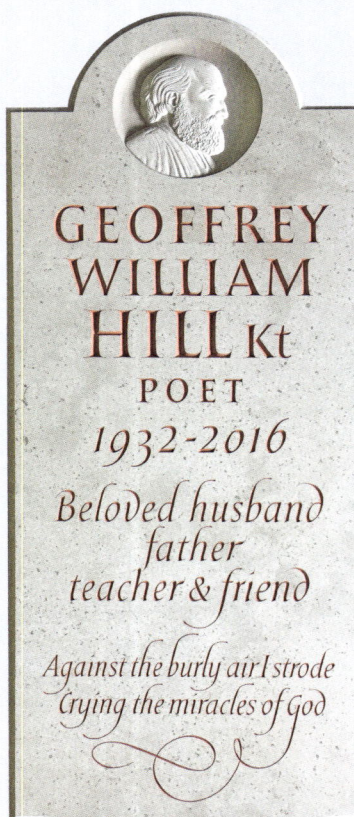

When carving smaller lettering the stone feels soft, almost buttery, with no major surprises such as big shells or air pockets. One can create a very fine, smooth edge thanks to the stone being so compact. Serifs and flourishes flow delicately as there is little risk of chipping or crumbling.

When carving a portrait in shallow relief one must use texture and shape to create the illusion of depth where there is very little. It is in this moment, especially in carving detail, that a consistent, dense stone becomes essential; all it would take is one big, loose shell in the wrong place and I might find myself having to re-carve half my work to absorb the loss.

ROXANNE KINDERSLEY

8 · REFLECTION

Lida: Although we each hold different perspectives, yours as a scientist and mine as an artist, I realise that the geologist and letter cutter are united in being passionate about rock, stone, creativity and exploring what we are doing on this Earth.

Steve: Both professions rely on the collegial atmosphere of small teams to deliver. Even in very large companies, with tens of thousands of employees, worthwhile work is generated in teams of ten or less. This lack of hierarchy enables a 'safe zone' of close working relationships and creativity.

Lida: I agree. The Workshop has always been limited to six or seven letter cutters. Each able to do everything, to step in at any point and to help out fellow workers. And there is a very direct personal relationship with our clients.

Steve: You and I have now both handed over to new leadership. I am grateful that the 'workshop' I led in the UK is still sustained within a global company context. Indeed, they asked me back for their most recent Christmas lunch with the invitation to *'see how they are continuing to flourish after your leadership set them on a great path'*. Heartwarming.

Lida: I have just gone through a similar process, having handed on the Workshop leadership to the next generation with Roxanne Kindersley at the helm, making certain that what is passed on is viable, exciting, well structured and has a place in society. I trust that it will move forward and follow the ideals by which my late husband David set up his Workshop in the 1930s.

Steve: In conclusion, I hope that, as a result of this book, our readers will experience a greater understanding and a greater affection for our Earth, and be inspired by the geological events that shaped it.

Lida: The dinosaurs left their bones as historical record; letter cutters hope to leave their inscriptions as evidence of people's thoughts and love set in a suitable stone.

The authors and the Workshop team. Top to bottom, left to right: Fiona Boyd, Emily Bunton, Paul Kindersley, Peter Miller, Rachel Iliffe, Vincent Kindersley, John Mawby Whistler, Debi Goodchild, Hallam Kindersley, Roxanne Kindersley, Beverly Moxon, Graham Beck, Steve Garrett, Lida Kindersley and Stuart Vallis.

Thanks

This book would not have been realised without the help and guidance of many people. We are very grateful to: Dr Nigel Woodcock at the University of Cambridge for the foreword and for valuable geological insights; Dr Robert Seidel at the Sedgwick Museum, University of Cambridge for hosting visits to the Watson Collection; Paul Johnson at the Geological Society of London for a literature search; Brenda Stone for expert editing of the evolving text; Kitty Hall for her companionship on field visits and proof reading; Graham Beck for his eagle eye on details in the text; Professor Thomas Sherwood for reading; and each member of the Workshop for their enthusiasm and engagement in the project, particularly Fiona Boyd for drawing the beautiful illustrations; and to Roxanne Kindersley for taking this book and the Workshop to new levels.

Geological Glossary

aragonite: *calcium carbonate formed by biological and chemical processes.*

calcite: *a stable form of calcium carbonate common in many limestones.*

Caledonian orogeny: *a mountain-building event (490–390 million years ago) when the Iapetus Ocean closed and Scotland and England were brought together in the process of continents colliding.*

Cambrian: *geological period (539–885 million year ago) during which the Welsh slates were originally deposited.*

Carboniferous: *geological period (359–259 million years ago) named after a key constituent of the Coal Measures and containing York stone.*

Cenozoic: *Earth's current geological era (66 million years ago to present).*

chalk: *soft, white, porous carbonate rock common in the Upper Cretaceous formations of southern England.*

chlorite: *a green mineral common in metamorphic rock, such as the Lake District green slates.*

Coal Measures: *Upper Carboniferous coal-bearing strata which powered the Industrial Revolution.*

Cretaceous: *geological period (145–66 million years ago) which ended with the mass extinction of the dinosaurs, caused by a very large meteorite.*

Devonian: *geological period (419–359 million years ago) in which land plants first appeared and during which Caithness Flags were deposited.*

diagenesis: *physical and chemical changes in sediments caused by water – rock interactions, microbial activity, and compaction after their deposition.*

dolomite: *a sedimentary rock that contains a high percentage of calcium magnesium carbonate.*

granite: *a coarse-grained igneous rock composed mostly of quartz, feldspar and mica minerals resulting from the slow cooling of magma deep in the Earth's crust.*

Iapetus Ocean: *seaway (600–400 million years ago) that separated ancient continents, revealed by fossil assemblages that differ greatly between Scotland and southern Britain, closed during the Caledonian orogeny along the Iapetus suture.*

igneous: *rock solidified from molten magma within the Earth's crust, such as granite; or lava on the Earth's surface, such as basalt.*

ignimbrite: *deposit of a hot suspension of particles and gases flowing rapidly from a volcano, known as pyroclastic flows, as seen in some Lake District green slates.*

iron pyrites: *iron sulphide mineral with the chemical formula FeS_2, also known as fool's gold, common in Welsh slates.*

Jurassic: *geological period (201–145 million years ago), beginning with a mass extinction event, and during which the Jurassic oolites were deposited.*

limestone: *sedimentary rock made largely of calcite, aragonite or dolomite.*

Magnesian limestone: *Permian rocks in north-east England made mostly of dolomite.*

metamorphic: *rock that has undergone transformation by heat, pressure and chemical alteration.*

Millstone grit: *Carboniferous coarse-grained sandstone found in northern England.*

Mineral: *a solid substance with a well-defined composition and crystal structure occurring naturally in pure form. Rocks are made up of one or more minerals.*

Moh's hardness scale: *ranging from 1 (talc) to 10 (diamond), ordering the harder materials that can scratch softer materials.*

oolites: *limestone made of small spherical grains formed of concentric layers of calcite or other minerals deposited in shallow seas.*

Ordovician: *geological period (443-488 million years ago) during which the Lake District green slates erupted.*

Permian: *geological period (299-251 million years ago) during which Magnesian limestone was deposited, ending with the largest mass extinction event on Earth caused by volcanic eruptions in Siberia.*

porosity: *void space in a rock, measured by the ratio of pore volume to the total volume.*

quartz: *mineral composed of silica (silicon dioxide), a key constituent of granite and sandstone.*

sandstone: *sedimentary rock deposited in water or deserts and later cemented together by minerals precipitated from groundwater, largely made of quartz sand grains (0.0625 to 2 mm in diameter).*

sedimentary: *rock made of the residue of weathering and erosion of existing rocks, or of detritus from dead aquatic organisms.*

shale: *sedimentary rock formed from mud, including clay minerals and quartz and calcite, splitting into thin layers (fissility).*

siltstone: *sedimentary rock with mineral grains (0.002-0.0625 mm in diameter) feeling grittier than shale when chewed and without shale's fissility.*

slate: *metamorphic rock derived from shale or volcanic ash with 'slaty cleavage' caused by clay regrowth in planes perpendicular to compression during events such as the Caledonian orogeny.*

tuff: *rock made of ash ejected from a vent during a volcanic eruption.*

Bibliography

BROOKS, C. (2017) A petrophysical, petrographic, mineralogical and geochemical investigation to determine the superior UK-produced roofing slate. Unpublished report, BSc Hons, University of Manchester. 78 pages.

DIJKSTRA, A. and HATCH, C. (2018) Mapping a hidden terrane boundary in the mantle lithosphere with lamprophyres. *Nature Communications.* DOI: 10.103 8/s41467-018-06253-7.

HACKMAN, G. (2014) *Stone to Build London: Portland's Legacy.* Folly Books. 311 pages.

HUGHES, T., HORAK, J., POULTRY, M. and COOPER, M. (2013) Portland Stone: A nomination for "Global Heritage Stone Resource" from the United Kingdom. *Episodes* Vol. 36, No. 3, pages 221–223.

HUGHES, T., HORAK, J., LOTT, G. and ROBERTS, D. (2016) Cambrian Age Welsh Slate: A Global Heritage Stone Resource from the United Kingdom. *Episodes* Vol. 39, No. 1, pages 45–51.

LOPES CARDOZO KINDERSLEY, L. and GAYFORD, M. (2003) *Apprenticeship: the Necessity of Learning by Doing.* Cardozo Kindersley Editions, Cambridge. 52 pages.

LOPES CARDOZO KINDERSLEY, L. and WAITHE, M. (2023) *Words Made Stone: the Craft and Philosophy of Letter Cutting.* Cardozo Kindersley Editions, Cambridge. 192 pages.

KINDERSLEY, D. and LOPES CARDOZO, L. (1981) *Letters Slate Cut: Workshop Philosophy and Practice in the Making of Letters: a Sequel.* Updated 1990 and 2004. Cardozo Kindersley Editions, Cambridge. 42 pages.

LARTIGUE, J. H. (2017) The textural, mineralogical and petrographic properties that define a high quality slate, focusing on a comparison of Penrhyn and Honister slate. Unpublished report, BSc Hons, University of Manchester. 50 pages.

KING, F. and LOPES CARDOZO KINDERSLEY, L. (2019) *Sundials: Cutting Time. The science and art of 27 Kindersley dials.* Cardozo Kindersley Editions, Cambridge. 139 pages.

MORGAN, N. (2016) Building stones explained 8: Gravestone geology. *Geology Today*, Vol. 32, No. 4, pages 154–159.

PURCELL D. (1967) *Cambridge Stone.* Faber and Faber, London. 115 pages.

THOMAS, I. (2005) Hopton Wood Stone – England's premier decorative stone. In: *England's Heritage in Stone*, Proceedings of a Conference, York, 15–17 March 2005. English Stone Forum. Pages 90–105.

WATSON, J. (1911) *British and foreign building stones: a descriptive catalogue of the specimens in the Sedgwick Museum, Cambridge.* Cambridge University Press, Cambridge. Reprinted 2015. 492 pages.

WINCHESTER, S. (2001) *The Map That Changed the World.* Penguin Books. 352 pages.

WOODCOCK, N. H. and FURNESS, E. N. (2021) Quantifying the History of Building Stone Use in a Heritage City: Cambridge, UK, 1040–2020. *Geoheritage*, Vol. 13, No. 12, 21 pages.

Dr STEVE GARRETT is a Fellow of the Geological Society, London, who graduated in geology at the University of Bristol in 1980 and in geophysics from Birmingham University in 1981. His first job was in Antarctic research. He then worked in diverse settings across other continents as part of the global energy industry for thirty years in Earth science, technology development and career management. Since 2018 he has focused on music as a guitarist, currently working on links between Earth science and the creative arts.

Dr LIDA LOPES CARDOZO KINDERSLEY MBE is an Honorary Fellow of Magdalene College, Cambridge. She studied graphic design at the Royal Academy in The Hague before joining David Kindersley in 1976 as an apprentice in his Workshop. She became partner in David Kindersley's Workshop in 1981 and joined him in designing and cutting letters, as well as training apprentices. After David's death in 1995 she continued the running of the Workshop with her second husband, Graham Beck, until 2023 when she handed over to Roxanne and Vincent Kindersley.

View the full list of Cardozo Kindersley publications and order online at:
www.kindersleyworkshop.co.uk/shop